Field with a view

Field with a view

Science and faith
in a time of climate change

Katharine M Preston

wild goose
publications

www.**iona**books.com

Overseas distribution:
Australia: Willow Connection Pty Ltd, Unit 4A, 3-9 Kenneth Road,
Manly Vale, NSW 2093
New Zealand: Pleroma, Higginson Street, Otane 4170, Central Hawkes Bay
Canada: Novalis/Bayard Publishing & Distribution, 10 Lower Spadina Ave.,
Suite 400, Toronto, Ontario M5V 2Z2

Printed by Bell & Bain, Thornliebank, Glasgow

To grandson William Shepard Constable,
Nana praying that the loons will always be there for him.

With gratitude to:

John, first and primary editor, sounding board, patient companion and, most of all, beloved partner

Our Iona Community Family Group,
who listen to and pray for me

Contents

Introduction

Sometime after giving up on the idea of being a ballet dancer or a cowboy, I dreamed of being a forest ranger. My reverie was explicit: I lived alone in a cabin deep in the middle of some national forest or park. My 'office' was a fire tower – overlooking thousands of acres of forestland.

The point was not particularly the job itself, the responsibility for watching and protecting the area. Rather, the essence of the dream was in the intimate knowledge that I would acquire about the place due to a long and observant relationship. It is March – I look for the sow bear to emerge from her winter quarters; I suspect there will be more than one cub, as the fall berry season was so prolific. It is May – there are fewer warblers than this time last year … I wonder why. It is October – I mourn the absence of the hulking bulk of the wolf pine, destroyed by lightning last July, that used to stand in solid green contrast to the colourful fall palette.

I *know* this place; I *feel* this place; I *am* this place.

Eventually, I persuaded myself I could make a vocation out of this dream. I could be assigned to nurture that intimacy in the name of a job. So I attended forestry school, obtained a degree, and then, as realities and personal relationships intervened, ended up predominately behind a desk for the rest of my professional career. Not in a fire tower.

The French mystic Bernard de Clairvaux said, '*Our yearnings shape our souls.*' I think he is right. And all our yearnings, fulfilled or not, are sacred.

Some of us yearn for intimacy with non-humans and with place. This yearning is an affliction of the modern world; in former times, people experienced that intimacy every day, eking out their existence alongside their fellow creatures. Over time, humans began to see themselves as separate, with a very specific and sacred role – appeasing and placating the gods, or God – which gradually placed them at the top of the hierarchy of the cosmology.

But ever since our very first view of planet Earth from space, we began to see ourselves differently, and finally, more realistically. This is frightening, challenging. Sometimes indescribably joyous.

* * *

I remember walking in the woods with my father or mother when I was very small, diligently looking for 'signs of spring'. There is an intimacy fostered by taking the time to *notice* the first funky spears of skunk cabbage; a precious relationship is established that sets a child on a journey. The journey does not necessarily have to be informed by scientific knowledge of the heavens, earth or humankind, but it helps to be *aware* of place and to acknowledge the yearning for intimacy with it.

Over time, I found myself fascinated by the question: Where do people place themselves in the *oikos*, the home, the household of the rest of the planet, and how is that reflected in how they live and in their concept of the sacred?

In college, I studied anthropology, in particular, the indigenous Hopi people of northeastern Arizona. I wondered how their rituals reflected their relationship to their harsh environment. My thesis was that without the rituals, they could not have maintained their existence on their marginal ancestral lands, where they have lived, continuously, for nearly one thousand years. In forestry school, I studied ecology, particularly human ecology, and, as it was the early '70s, became aware of the often-negative human influence on natural systems. Why this disconnect? What did it say about how humans saw themselves in relation to the rest of the natural world? In seminary – some thirty years later – I explored how the God/human relationship and religious teachings, particularly as reflected in progressive Christianity, liberation and process theology, might mend the human/environment relationship.

Looking back now, the forest ranger dream was my search for grounding. I think I was seeking confirmation that a human being could indeed learn to live in close harmony with a small bit of the planet.

For a long time, I resisted writing about climate change. I wanted to write lyrical descriptions about the landscapes surrounding me. I wanted to rest in the here and now, in the moment, in this place. I evaded the issue, pushing the terrifying science to the margins of my mind, along with the increasing evidence that migrations and wars reported on the news were directly or indirectly related to local disruptions due to a changing climate.

But the evidence caught up with me when I realised that some of the most precious beings and landscapes around me were already changing. Scientists were beginning to hint that we might already be beyond the 'tipping point' of catastrophic change. I look at my grandchildren. What kind of a world will they inherit and how will they inhabit it? And then there was the irrefutable fact that the people most innocent of contributing to the problem were the ones most affected and least able to adapt.

I simply could not ignore the injustice of this.

My scientific and theological training insistently whirled around in my mind, forcing me to consider some existential questions.

How do I, as a rational person of faith, make sense of climate change? I don't mean trying to understand what happens, what might cause it, or how it affects humans and non-humans, although as a member of the species *Homo sapiens* I embrace the wisdom of trying to find out these things. But how do I, how do we, make *sense* of it? How do we incorporate this new reality into our lives?

Climate change forces people of faith to face some very profound and challenging questions about the God/human relationship:

> *How can God let natural occurrences such as hurricanes and floods and wildfires hurt so many innocent people?*

> *Would God create a human species so flawed that we could do this to ourselves?*

> *Would/could God actually let the human species die off?*

And for all people, with or without faith in God:

> *What are our responsibilities to the people suffering because of climate change?*

> *What are our responsibilities to the rest of creation?*

The reflections that follow are formed by a lifetime of loving the intricacies and wonders of a planet that never ceases to awe, inspire and comfort me. Most particularly, the reflections spring from the pinewoods of my youth in Massachusetts, the fields surrounding our farm in the Champlain Valley of New York and the contiguous Adirondack Mountain wilderness. And they spring from my observations of the hopeful human/earth relationships developing in our small rural community. These reflections are also formed by my faith, which sees the earth as a sacred manifestation of God, and its human and non-human inhabitants as neighbours to be loved and defended from the injustice of climate change.

At the moment, I see people with different arms tied behind their backs, trying to save a world in crisis. There are those who have abandoned faith, because they think God is the same thing as church. A relationship is thrown away with the institutional bathwater. There are others who have abandoned science, because they feel it threatens a biblical narrative and somehow negates the workings of the Spirit. The possibility of awe and wonder at scientific discoveries is thrown out by a narrow definition of 'truth' and of the miraculous. I see both positions as sadly shortsighted.

So these reflections strive to be both theologically challenging and ecologically informed. I hope for readers with open minds: scientists leery of faith but open to unanswered mysteries, as well as believers who see value in every miraculous scientific discovery and are not afraid to say so. Many of my friends take the 'spiritual but not religious' road. I hope that they will see value in some of the unconventional views of divinity and church that I present.

I relate an ongoing journey; my personal response to what I believe is an apocalyptic moment. I have moments of anguish, moments of unbridled fear, but also moments of joy. Frankly, the hope is harder to come by these days, perhaps because I am discriminating: I do not want to embrace cheap hope (humans have always come through) any more than I want to accept cheap anguish (we are doomed). Solutions, if they exist, are far more complicated and nuanced. Thinking about these things has sharpened my relationship with God. I do not have answers to all the questions posed; I can only relate what I feel and what I have learned, what decisions I have made in response. I invite you to ponder the questions, journeying alongside.

No inconsistency

> *Less and less do I see any difference*
> *between research and adoration.*
>
> Teilhard de Chardin

I am deep in the mountains, alongside a lake. I see mostly stillness, but also the gentle, almost imperceptible movement of a layer of mist gliding across the lake's quiet surface. From my vantage point on the western shore, the mist moves slowly toward the opposite ridge, eastward and southward, where, this late in the season, the sun will soon make its midmorning appearance from behind a forested slope. Although the scene is replicated on the lake's mirror surface, the thin blanket of mist obscures the line between substance and image. The peaks of two or three giant white pines float incongruously in the mist, and where the mist begins to dissipate higher up the forested slope, there is a hint of autumnal colour.

A squirrel pitches a warning chatter into the stillness, and a raven sounds its washboard croak.

There is a suspension of time and breath for a moment, and then the ridgeline begins to glow, shimmering with the sun's rays at the season's designated azimuth. The pace of movement quickens. The mist accelerates across the lake's surface, rising to meet the warming air. And as the sun's rays finally spill over the ridgeline, a joyful greeting-dance begins, mist and sun twirling above the surface of the lake.

The dance can be explained through basic physics: the interacting properties of air and water and temperature gradients. It is not a one-time event; it has happened millions of mornings before, right here on this wilderness lake, and will happen millions of mornings hereafter, whether I am here or not. But the event's scientific explanation does not reduce the grace of the moment, and the revelatory nature of my experience of it does not deny the science. The two aspects blend together nicely in my soul. Indeed, they enhance each other.

Physics/God. Ecology/theology. I simply do not see the incongruity. Most people of faith I know fully accept Darwin's theory of evolution and other revelations derived through sound scientific enquiry. Trouble is, even the more liberal mainline churches rarely *talk* about science. I have a notion that some people in the pews may secretly worry that they may be 'less faithful' for embracing a scientific view of the world. And some scientists remain fearful that reason will somehow be fatally contaminated by faith.

I am weary of qualifying my words when I write. If I express my awe and wonder at the beauty of Creation-with-a-large-'C', some will assume I am a fundamentalist who does not believe in evolution. Even a small-'c'-creation is suspect, as it implies something formed from nothing by 'something'. Nature-with-a-capital-'N' invokes Transcendentalism. As much as I admire Thoreau and his close observation and love for his surroundings, his 'Nature' is decidedly other-than-human, as is nature-with-a-small-'n', for that matter. The general use of the word implies mountains, frogs, lions … but *not* human beings. A dangerous exclusion.

I have not come up with a solution.

I agree with scientist/philosopher Teilhard de Chardin about research and adoration. For me, trained in both science and theology, there is a very fine line between the passion of faith and the passion, awe and wonder evoked through a scientific understanding of and relationship with things wild.

After all, science is miraculous. An unsurpassed collective endeavour, science faithfully builds upon itself, each generation refining the work of the preceding one. As an international language, it brings the world together, bridging culture in a way that diplomatic endeavours envy. Scientific revelations are like a never-ending Advent calendar: we open one door at a time to reveal something more of the picture behind, but never all, for what is invariably revealed are more doors into as-yet-not-conceived-of realms. The scientific revelations are gifts – diligently worked for, but still gifts. Most scientists would admit to the presence of at least some serendipity in the process of scientific discovery.

For sure, science is not always 'right' in the sense of 'correct'. Mistakes are often what propel research onward. Nor is science always authoritative, because there are other ways of knowing than through our minds. After all, we are described as *Homo 'sapiens'* not *Homo 'ratio'*. Few would equate wisdom exclusively with knowledge.

Pope Francis, in *Laudato Si*, says that science and faith must remain in dialogue with each other: '*It cannot be maintained that empirical science provides a complete explanation of life, the interplay of all creatures and the whole of reality.* This would be to breach the limits imposed by its own methodology' (my emphasis). Ethical principles do not arise in the abstract, he notes. '*Nor does the fact that they may be couched in religious language detract from their value in public debate.*'

The media doesn't help the discussion between science and faith. Because it makes for better copy, the media gravitates towards sensationalism, which sometimes rests on the margins of intellectual thought. I once saw a 'religious' person interviewed on television who said, literally laughing at the absurdity of the notion, that there was *no way we are related to apes!*, given the biblical account of creation. I find this kind of talk in the 21st century bewildering. A mindful church should be on this sort of thing in a flash – because it is bad theology and a tragic misinterpretation of a book with great wisdom. This kind of thought is particularly unfortunate from a pastoral point of view because it wreaks havoc on the psyches of people who are both reasonably faithful and faithfully reasonable. And it sure doesn't encourage young people to remain in the pews. For them, rejecting established science is simply silly.

The church has spent a great deal of time convincing itself, and us, that because we think, and because we have knowledge of good and evil, we matter. And we *do* matter, to each other as humans, of course, and for some people, also to God. But in ecological systems, the community is always paramount and, our deportment to the contrary notwithstanding, humans are inhabitants of a much larger community.

Climate change seems to broaden the chasm between science and faith: it has become easier for some to deny the science than to face the existential questions prompted by a vision of a world radically changed by the way some of us live. Yet, the two have never needed each other more. Writer and environmental activist Bill McKibben has commented that because of the moral compass they provide, religious communities are deeply important to the efforts to mitigate the effects of climate change. This wisdom is not new: some decades ago, ecologist Aldo Leopold argued for the integration of a scientifically based land ethic into the moral code of society, suggesting that much could be gained by religion's adopting an active role in that integration.

The scientific environmental community increasingly recognises the church's capacity to shape worldviews and provide the moral authority indispensable to the cause of scientific enquiry. And faith communities use ecological statistics and scientific rationale to reinforce their efforts to preserve God's creation, as well as to promote justice for the poor, who are suffering on the front lines of human-induced destruction of the environment.

Not a moment too soon. We need all the cooperation we can stand. Arguing about our ape ancestry will do nothing for our descendants. We've got to get beyond the false conflict between science and religion for it takes our eye off the real matter of trying to mitigate the effects of climate change. Once again, we rearrange the deck chairs of the religion versus science debate while the iceberg of global climate disruption looms across our bow.

Despite what the tabloids say, there will be no end to science or to God (or gods) until the human species ceases to exist. And even then, some primate cousin may be well on its way to considering *how* that apple just fell on its head, and even *why*.

The environmental movement was born of people's new understandings of the ecological sciences as well as from deep spiritual experiences of the natural world, like mine this morning, watching the mist rise from the lake. The shared experience is a powerful incentive to act.

Wild apple trees
and Process theology

Religion will not gain its old power
until it can face change
in the same spirit as science does.

Alfred North Whitehead

Around the second year after moving to Wild Orchard Farm, we noticed something extraordinary – there were dozens of what appeared to be wild apple trees spread throughout what we call the pasture lot, 46 acres of uncultivated land that is no longer open field. 'Wild' was a guess; the trees do not appear to be planted in any cogent fashion, and are tall enough that harvesting apples would require a specialised ladder, tall and tapered, to fit up into the branches. The story of how excited we were during those first years, cultivating visions of well-pruned, healthy trees producing multiple distinctive varieties for eating, sauce, cider and drying, is another story. Suffice to say that the reality of the work and the capriciousness of trees that had for decades been totally on their own, to do as they pleased without human interference, made planting more predictable lettuce in an orderly kitchen garden near the house much more appealing.

When it is no longer ploughed or grazed, the landscape in this region returns to forest with amazing tenacity and speed. So in this abandoned pasture lot, the mature apple trees are surrounded by many cedars, juniper bushes and some white pines, all early successional species. 'Ecological succession' is simply the replacement of one community with another. It is not a particularly orderly process, although the conditions of the soil, forest cover and water will help determine what grows where.

How did the apple trees get there? At one point, there was a small orchard just to the south, so I suspect the randomness of the pasture-lot apple trees was simply due to animals and birds distributing the seeds, through their gut or by messy eating.

When I was in forestry school, we were taught about 'climax communities': populations of animals and plants best suited to the ecological conditions of a particular area that remain relatively stable after a series of successional communities. But recently, ecologists have backed off a bit from using the word 'climax'. There is too much finality implied. Instead, there is no self-perpetuating *stable* biotic community. Rather, it is simply a *relatively* stable holding pattern until something else occurs: drought, fire, disease, hurricane, particularly frigid winter temperatures or myriad human interferences. Ecological communities are dynamic, constantly changing, adjusting.

In divinity school, I discovered an elegant theological explanation that matched the innate dynamism of ecological systems.

Process theology is a fundamentally key concept for me because it attempts to provide plausible explanations for how and why things are the way they are that are consistent with both scientific observations of the world and the human experience of the divine.

Developed by American philosopher Charles Hartshorne, whose life literally spanned the entire 20th century, and American theologian John Cobb, co-founder and director of the Center for Process Studies in California, process theology is a theological understanding of the philosophy of Alfred North Whitehead, an English mathematician who lived from 1861 to 1947. Process thought arose in conjunction with the revolution in science itself: the acceptance that there is nothing static or mechanistic about the universe. The basis of reality is nothing stable at all – but all movement, change, dynamism, chaos.

Process thought sees realities as processes or events, not as things. It emphasises becoming and relatedness as categories for understanding the world rather than substance or being. Becoming is more basic than being.

Our empirical observations of the universe support this theory, from the structure of the atom (we cannot say where the electrons actually are at any moment yet we can predict where they might be at some time), to ecological principles (all individuals are intrinsically part of communities and cannot exist outside of one), to evolution (life seems to evolve in a particular direction, influenced by both surroundings and genetic change), to, on a far more macrocosmic scale, the expanding universe. Reality is in constant flux, displaying what might be called an 'emergent state of being', always in relationship.

All this processing could be random, but again, our empirical and experiential observations indicate that there is an intrinsic order in the chaos, and the order seems to proceed in a particular direction.

But is God involved? some ask, including me. And if so, how?

Process theology uses the concept of *panentheism* to explain how God is related to and is affected by what happens in the world. Panentheism (all is in God) is differentiated from pantheism (all is God). In Charles Hartshorne's theology, God is not identical with the world, but God is also not completely independent from the world. God has a self-identity that transcends, but the world is also contained within God. God experiences everything there is to experience. A rough analogy might be the relationship between a mother and a fetus. The mother has her own identity and is different from the fetus, yet is intimately connected.

Long ago I rejected the conclusion that the theory of evolution somehow negated the existence of God. That conclusion seems ridiculous to me. There is nothing 'unnatural' about God, so why couldn't God be involved in the process of natural selection, succession and all the other processes that make the world work. I also did not believe in predestination, for humans or anything else on earth. It seemed unnaturally callous and, frankly, pretty boring that God had some plan, a rigidly predetermined sequence, proceeding according to some divine timetable for this magnificent world. *There had to be room for change.*

In ecological systems, we see a world of constant adjustment, the dynamic interplay of parts: sun, grass, voles, human beings, streams, mountains. In process theology, the innate nature of the divine force elicits that constant adjustment and, indeed, is part of it. For people of faith, this is consonant with our intuitions of grace: God is ever active in the world – enticing the world toward better possibilities, providing a future out of every past – no matter how inadequate that past. But this is important: there is no predetermined path, no coercion and absolutely no punishment if we (or for that matter, if voles) do not follow God's enticement. God is simply a supreme persuasive power. There may be consequences: a vole 'choosing' to flaunt itself in the middle of a field with a hungry hawk overhead will face death. But it is not a punishment.

The 'beings' of the world, from atoms to humans, have their own freedom of action. This freedom belongs to them by nature. Their freedom constitutes a *limitation upon God*. And because God is 'in all things', *God is changed* by the actions chosen.

This is a very different model from the traditional view of God as fully omnipotent, unchanging and judgemental. For some, it may be just too big a step from what many people of faith have accepted for millennia. But for me, it is crucially liberating, because it not only explains 'how things work', but helps me answer the question of theodicy – bad things happening to good people. God is unpredictable because *God does not have the power of full intervention*.

A somewhat crude example might be seen in what we call natural disasters.

Physical laws dictate that after aeons of force building up in a particular direction, there can be a shifting of tectonic plates, causing an earthquake. But the disaster part is the result of choices made by humans. Despite the constant enticement by God for us to seek better understanding, acknowledgement and acceptance of the necessity of living within certain ecological, sociological, not to mention geological, constraints, the wrong course is chosen, erring on the side of our human preferences.

In some cases, we might make these choices. But the poor don't have this freedom. And as disaster after disaster has shown, it is almost always the poor who are hurt most, as they live in the most marginal housing, not by choice, but by necessity. Building occurs on steep slopes along fault lines with inadequate materials. Or houses are built on the flood-plains of rivers, on barrier beaches, or on slopes denuded by logging or fire. Our place within delicate ecological systems is ignored and there are consequences.

This is not an indication that God doesn't care: the entice-ments toward better decisions are there all along the way from the very beginning. And when the consequences of our bad decisions appear, God does not disappear: God is in every victim, first responder and volunteer helper. God is right alongside.

I am reminded of the story/parable about the man who dies from flooding. He challenges God in heaven: 'Why didn't you save me?' And God says to him: 'I tried! What do you think the two boats and the helicopter that came by were trying to do? You sent them away proclaiming, "No need! God will save me."'

Although process theology is profoundly theocentric, the roles of Jesus and the Holy Spirit are not ignored. Process theologian Bruce Epperly describes Jesus as *'a reflection of God's aim toward creative transformation, calling humankind forward from what is to what can become.'* Jesus is *'an energetic field of force'* – for healing, for companionship, for love, for hope, essentially the prime example of divine persuasion in human form. In a sense, Jesus is a reflection of the divine so true as to be indistinguishable from the source.

To me, the concept of panentheism (God within all) is simply another way to understand the workings of the Holy Spirit.

Back to the wild apple trees.

I see in the pasture lot a metaphor for process theology. The succession is seemingly random. An apple seed falls here, instead of there. But many, many things determine whether it will live or die, and whether it will thrive into the future: the DNA within its cells, the conditions of soil where it falls, the interference from devouring deer, an enthusiastic farm owner, or lightning strikes. The seed, its sprout, its maturity – and even its death, returning to earth – is always a 'becoming' thing. God is that process.

I find great solace in this less-than-all-powerful explanation of how God works. It is consistent with my understanding of ecological systems and my faith in the alongsideness of God. It reminds me that the choices I make are important, and that the reverberations from those choices ripple outward in ways that I (and God) cannot possibly predict. So I had better listen, watch, notice, practise empathy, and love. Right alongside.

Imago Dei

> *To see a world in a grain of sand*
> *And a Heaven in a wildflower,*
> *Hold infinity in the palm of your hand*
> *and eternity in an hour.*

William Blake

In the pause time of late winter, my eyes scan across the dullness of the farm fields. There is a slight hue of green offered by that portion of the field cut and trampled most the season before: our narrow footpath winding its way to the pasture lot. A memory of previous seasons is stirred in my winter-numbed brain, an anticipation of maybe a shock of field daisies here, scattered black-eyed Susans there.

Then, one day near the end of March, during a moment of sunshiny respite from freezing rain and snow, a brief preview of coming attractions: one or two tree swallows appear, squeeze in a few graceful soars, and then disappear from the farm. A week or so later, with the moist air warmed to a more delectable temperature, other swallows arrive – and suddenly the air is full of them.

It is as if the air itself has burst with joyful recognition and welcome of spring. My spirit soars.

It is extraordinary how the arrival of a small bird, flying into one's view, shifts the entire seasonal perspective from 'end' to 'beginning'. If swallows – then surely, daffodils, lilacs, even spring peepers in the pond, soon!

Between soars, the swallows line up, carefully spaced along the telephone wire, catching up on the gossip. After a break, they swoop and grab soft brown blades of last season's grass and pine needles, the occasional stray feather, and busily disappear into the nest boxes.

I remember reading in some ornithological source how a group of swallows will chase a feather in flight, carrying it high, dropping and retrieving it as it floats in the air, in an aerial competition with one another. I found this hard to believe. But then one spring, outside our window, I saw it happen. I stood transfixed; the dancing reality took my breath away. To me, this is a moment of unexpected grace, and I say to myself: 'This is Holy Spirit stuff': my way of acknowledging the grace and saying thanks.

The writers of scripture traditionally characterised the Holy Spirit as a hovering dove. As I watch the swallows, I wonder why. Unless those in the Middle East are different from ours, doves don't actually *hover*; they always seem rather frantic and fearful, their wingbeats noisy, rapid. Besides, hovering implies a pause, a suspension of activity – an unlikely option, I think, for the Holy Spirit.

I can't describe what the Holy Spirit is – but I am pretty good at recognising what the Holy Spirit *does*. I see results: what the Holy Spirit *causes*, or *inspires*, or *provokes* or *activates*.

There is a wonderful conversation in the book of Exodus between God and Moses that illustrates this limited disclosure. Moses asks to see God's full glory and splendour, but is told that witnessing it will kill him. Instead, God suggests that Moses hides in the cleft of a rock and be covered by God's hand. *'I will make all my goodness pass before you,'* says God. Then God will remove the hand and Moses will see God's back, but not God's face: Moses will see the *results* of God passing by (Exodus 33:17–23).

The wind blows. You do not see it, but you see the results. Breezes ripple the water: the grasses dance. Kind of like that. After all, the word for Spirit in Hebrew (*Ruah*) and Greek (*Pneuma*) also means wind or breath. The latter is gender neutral, but the former is feminine. This is important to me. Acknowledging the Spirit as feminine reminds us of the mothering care and proud sisterhood present in the fullness of God.

Most explanations relating to the Trinity, the belief that there are three 'persons' of God, Son and Holy Spirit, are not relevant to me. Confined by a dark grove of doctrinal or dogmatic answers to questions, many of which were never meant to have final answers, the Spirit withers. Our challenge is to expand our viewscape to encompass as much of the landscape of possibilities as we can. She needs room to soar, as in an open field. Swallows seem to know how to do this well.

I prefer the image of the Spirit as swallow – soaring with unrestrained diligence in grace and joy.

I think about other images: The writer of Genesis declares that humans are '*made in God's image*', seemingly restricting the attribute to just humans. Many take this far too literally; I suspect it was meant as a metaphor, perhaps to sooth insatiable human egos. It has become a hook upon which to hang our self-importance.

For my faith, it is simply *too limiting* to contemplate the divine image as being revealed in just one part of creation.

Add swallows. Also koala bears, mountains and lupine. Why is the human exclusivity assumed? Is there any passage in scripture that states we are the *only* ones made in the image of God?

Some tell us humans are in 'God's image' because of our capacity for *self-reflection*. This confuses me. It may help us deal with *our* world, but it wouldn't help, for example, a fourth generation monarch butterfly find its way back to Mexico from Canada, geographically retracing the extraordinary genetic relay of its ancestors. Inherited memory of migration routes may not be particularly useful to humans. Why should one genetic attribute be intrinsically more pleasing to God than another?

Others argue that to be in the image of God, one must be *in relationship with* God. Agreed. But aren't we all? To qualify God's relationship with a polar bear or a newt as being somehow inferior to that of God's relationship with a human, and therefore not revealing of God's image, seems hubristic and self-serving. I know nothing about the nature of the relationship of a newt or polar bear with the divine.

Evolutionary science levels the playing field. As human beings, we are but a recent twig on the tree of evolution, not anyone's end point. There is nothing that makes us intrinsically more 'valuable' than any other species. Our success, after all, has yet to be tested in the crucible of time. (Microbes and cockroaches have sustained themselves a lot longer.) Indications at the moment are that our big brains may not be as reliable and sustainable an adaptation as we had hoped. Nomenclature of ourselves as *Homo sapiens*, the 'wise' member of the genus, just might be too optimistic given the state of the planet we voraciously inhabit.

We need to move beyond this attitude of exclusivity of image. One God, many images. Panentheism: 'all things in God', 'God in all things'. This rejects the traditional dualism between sacred and profane, between sentience and non-sentience, and just as surely between human and nature; it also implies that *everything* contains an image of God. Yes – this is very hard to accept for many of us with our hard encapsulation of human selfhood. But we must try: for the sake of the planet and for the sake of our souls.

For now, I make sure the telephone line stays strung in the air for the swallows' perching. I scatter a few downy feathers that have conveniently exited the old sofa cushions. And then I watch the joyful soaring of Spirit.

Earth excepting humans?

Above all we should, in the century since Darwin,
have come to know that man, while now captain of the
adventuring ship, is hardly the sole object of its quest,
and that prior assumptions to this effect arose from the
simple necessity of whistling in the dark.

Aldo Leopold

Deep in the high mountains of the Adirondacks in New York is a lake the dark colour of rich earth. Sometimes the surface is as smooth as velvet, but more often it ripples, streaked with dancing lines of silver. The rounded peaks of the eastern ridgeline high above the lake bow gently northward in homage to their sculptor, the retreated Laurentian glacier. On the slopes above the lakeshore, a few giant white pines, generational remnants of more abundant times, emerge above the blanket of spruce, balsam and northern hardwoods.

This morning, I sit on a dock on the western shore of the lake. Behind me is Sprucewood Camp, built by my grandfather one hundred years ago as a rustic retreat from his New York City toil. He came here to fish, to spend hours talking to his local guide, and to read at night by the light of a kerosene lamp.

I like to think he also spent time just sitting on this dock, pondering things, as I am now.

At this camp, tradition reigns. There are few apologies for not moving forward with the changing times. For some members of my family, the smallest changes have caused heartache, while for a few others the slow pace of change is a constant frustration. But in the end, all cherish the inheritance with unabashed joy. Prayers are whispered for the grace of heart to preserve the camp and its traditions for generations to come, with an accompanying resolve to do just that.

I am conscious of and thankful for the history and constancy of our camp. But that constancy is jeopardised now. Unlike my grandfather, I cannot be sure that the view from this dock will remain the same for my grandchildren.

We are told that here in the North Country, climatic changes will be felt sooner and with greater magnitude, particularly in the wintertime. Many local plant and animal species rely on the dormancy period provided by genuinely cold winters and a snow cover lasting at least a couple of months.

What might be lost? Things that I cherish: the stately spruce trees for which this camp was named; the fern-like tamaracks scattered around bogs; sugar maples with brilliant orange colour in the fall; balsam fir, the boughs of which line our lean-to in a soft, aromatic mattress. Moose, marten, mink frogs. Alpine tundra remnants that still grace some of the highest mountains.

The boreal and cool-temperate forests of these mountains will not disappear overnight; but during the next century, climate change will chase the forest community surrounding this lake northward, well into Canada. Southern species will migrate north to take their place, and the shores of this lake will gradually evolve to resemble those now in North Carolina or Georgia.

No offence to our southern neighbours – but I am selfishly happy that I will not live to see the change.

'Within a century,' say the scientists. I used to think that one hundred years was a long time. But having lived 70% of a century changes one's perspective. Only yesterday my grandfather sat on this dock. Tomorrow my grandchildren's children will be here. I pray that they may all experience the same breath-catching thrill I feel now as a loon surfaces from its dive close enough for me to see the drops of water roll off its back ...

Loons. There are few sounds more spectral and more indicative of wildness than the call of a loon, and for six decades those calls have conjured Sprucewood in my heart. I weep at even the *possibility* of no loons on this lake.

It is in settings such as this one – on this dock, alongside this lake, in this set of mountains, on the face of this earth – that the human animal (at least this one) contemplates existential questions. My family, even 'to the seventh generation', is only a brief sojourner on this ancient lake. My personal experience at any precise moment in time is an instance of no importance whatsoever to the place itself. The water changes colours as it always has, grey-blue to green-brown, and ripples from silver to black. The sun, moon, stars rise and set, deer come and go, loons call, Great blue herons stalk, squirrels and chipmunks chatter. Generations of these continue to come and go, and it has *nothing to do with me.*

But there is no doubt the human species has changed the face of the earth. Many scientists are calling for the designation of a new epoch, the 'Anthropocene', since human domination of biological, chemical, and even geological, processes on earth is a reality. Our dominance is indeed epic in the sense of how long the consequences of our present actions will carry into the future.

Atmospheric chemist Paul J. Crutzen and journalist Christian Schwägerl:

> *A long-held religious and philosophical idea – humans as the masters of planet Earth – has turned into a stark reality. What we do now already affects the planet of the year 3000 or even 50,000 ... It's no longer us against 'Nature'. Instead, it's we who decide what nature is and what it will be.*

I agree with this conclusion but take issue with its implied removal of the human species from 'Nature' or even 'nature'. As long as we view ourselves as separate from, and superior to, nature, the reality of our intimately shared existence will escape our consciousness.

Over time, humans developed larger brains to imagine and invent, and freer hands to fashion technological responses to the challenges of daily living. We also developed self-reflection and consciousness of our own mortality, two attributes which may or may not prove to be advantageous over time. After all, we haven't been around very long.

Recently, the human brain appears to be displaying some disturbingly self-destructive tendencies. I seriously wonder if human beings are just a passing phase or even an aberration in God's overall intention.

Perhaps not surprisingly, our theologies – that is, our religious 'truths' and explanations, monotheistic or polytheistic – have perpetuated the concept that human beings are somehow *special* to God, and more important than any other part of creation. Despite millennia of study, scriptural interpretation and creedal assertions, no proof of such an idea is forthcoming.

To me, God within the whirlwind described in Chapter 38 of the Book of Job implies not that God is immune to the sufferings of Job, but that there are *just a lot of other things going on.* 'Where were you when I laid the earth's foundation?' bellows God (Job 38:4). This is followed by two full chapters describing in gorgeous and minute detail all parts of creation that have nothing at all to do with humankind.

'Whistling in the dark', indeed.

Process theologians state that God's ultimate creative aim is toward *beauty* as well as what they call *complexity*. I agree that God's creative love seeks beauty. But I am not so sure about complexity, because for some people, it would follow that God loves humans more because we are 'complicated', that our genetic make-up is further along some imaginary value-added evolutionary scale.

In the end, the process of evolution is value-neutral: there are no favourites. If a person of faith truly accepts the science of evolution, then the face of an ant must be as cherished by God as the face of a child.

Those who insist that humans are totally and somehow ultimately significant ('See how we can change the planetary climate!') forget that *we do not know the end of the story*. For the time being, the planet adjusts: ice melts, sea levels rise, ever more violent storms dissipate the heat. These are the planet's counter-reactions to carbon overload; it has happened before and will happen again. And such 'adjustments', like evolution itself, are completely value-neutral. But it is the runaway *pace* of earthly changes with their origin in wilful human behaviour that is ethically untenable, when some of us are in a far better position than others to weather such changes.

So let's go a step further: as the more dire of the climate disturbances play out, might not our species as a whole find itself dangerously close to extinction?

There is a lot of disagreement about this. 'We'll adapt,' say some, 'always have.' Yes, so far, but our 'so far' is a minuscule instant in geologic time. 'The earth has gone through this before,' claim the skeptics. But they forget that *humans were not around* in that 'before'. Our species has never had to endure these climatic disturbances.

Although we feel very special to one another (and to our dogs), our species, like any other, is not *ultimately* indispensable to the ecosystems of which it is a part. Even those ecosystems manipulated beyond recognition to our advantage would revert very nicely to some other viable state were we to disappear tomorrow.

It is hard for us to get a grip on the idea that Earth would proceed without human occupation. Agnostics, atheists, people of faith, and even those-who-would-rather-not-think-about-such-things-thank-you-very-much, all share an understandable reluctance to contemplate that, while some species may endure, *Homo sapiens* may not be one of them. We seem always to assume that there will be *someone* like us around to see and reflect upon the apocalypse.

Whew! What does it do to the human psyche to face this possibility of termination? No wonder there is a burgeoning field of ecopsychology. Speaking about climate change, British psychotherapist Mary-Jayne Rust tells us: *'Even though we are waking up in a big way to the crisis, there is still a great deal of numbing, apathy and denial. We are having great difficulty in making even the simplest of changes to our lives … Overwhelming guilt about the damage we have done can block our thinking, and make us very defensive.'*

Amen. One can begin to understand the incomprehensibility of climate change among its deniers. Facing the next step – the possibility of extinction of the species – is just something we do not want to see on our radar screens.

'But wait,' says the person of faith. 'Won't God save us from ourselves?'

I am not at all sure that God can. In the context of catastrophes like tsunamis and hurricanes, or human atrocities such as genocide, slavery and the rape of children, divine omnipotence does not make sense to me, at least not on the part of a loving God. Process theology counters theodicy (*'the vindication of divine goodness and providence in view of the existence of evil'*) by saying that God does not coerce and intervene so much as *persuade*. God is not omnipotent, but rather is self-limiting. Involved, yes, and always present, but not directly interfering. Instead, God works alongside us, inviting, offering, persuading all of us (including whales and bacteria) toward what is best, what is beautiful, what complements the whole.

God can't save us from ourselves without our help. God doesn't fix. God accompanies.

But then I am compelled to ask: Will God accompany us even toward our self-annihilation?

In the Hebrew Bible, God is often depicted as angry, revengeful, punishing. Thus: *'The fear of God is the beginning of wisdom.'* Perhaps the fear should not be of God's anger or disapproval but of the grace-filled truth of God's constancy, God's reliable love for all of creation, *irrespective of the human condition.* *'Humankind is embedded in the process of planetary evolution,'* says process theologian Bruce Epperly. If humans choose to self-destruct, God will accompany that – and will continue to nurture other forms of evolution.

It may be harder for some to love a God that does not love human beings at the exclusion of others. I love human beings, but I love God more and trust God more. I don't mind that it is not just about us. For me, the simple *right-now-existence* of this silver-streaked lake, these bowing ridges, these towering pines, the loon swimming close enough now for me to look into its wary red eye, is enough to make me want to melt in faith and gratitude and love. There is an awe-filled completeness, with or without me, my family, and even my entire species.

I guess the challenge to me is to accompany God, regardless.

Grief

> *We have met the enemy and he is us.*
>
> Pogo

Early this morning, as I walk out to retrieve the newspaper, three loosely aligned skeins of geese fly roof-top-low over the farmhouse and the maple tree – blazing yellow. There is some merging of the tangled lines, and much noisy discussion about it all. The mid-autumn sun, already arched well to the south, illuminates their rhythmically undulating bodies – so very close! – against the blue open sky.

It is breathtaking: a moment of grace.

In one of her poems, Mary Oliver named her house 'Gratitude', in response to her close and precious experience of the natural world around her. *Notice* is surely one of the first things God invites us to do, but for me, at least, and apparently for Oliver, attentiveness is not the end of it. Geese-moments and other sublime experiences of the wild are an unexpected gift: after the 'wow' comes an overwhelming urge to say 'thank you'.

At these times, I thank the transcendent presence I call God. Perhaps it would be the same thing to thank fate or circumstance for putting me in the paths of flying geese. I don't know. Thanking God works for me.

As I watch the fleet of geese move on over the fields, their random honking fading and their bodies becoming a wobbly dotted V to the south, another reality interrupts. I recall the comment by one climate scientist that we are over-carbonising our way to an *'entirely different planet'*.

Heart-stricken, I hear myself cry: 'I don't *want* an entirely different planet!'

Granted, I cry from a position of privilege. I have never had to worry about a typhoon taking the roof off my house or the rains arriving in time to save the parched crops that will feed my family and me. I have lived and travelled in places where the beauty and sublimity of the landscape left me speechless with wonder and gratitude. Naturally, I hope my grandchildren and great-grandchildren 'to the seventh generation' will experience the same security, and be able to hear goose music floating down from the sky.

But alongside my selfish whining in privilege is my awareness that the evolution towards an entirely different planet is already causing acute suffering for millions of people. They are unable to adapt to rapid climatic changes because they are poor, marginalised and voiceless. They are desperate simply to live where they have always lived, but instead are forced to flee drought or flood or the all-too-human wars that ensue, becoming the world's climate refugees.

Climate change is the greatest existential and moral issue of our time, perhaps of any time. To many people and places, extreme weather events have already brought destruction, loss of livelihood and death. To others, the prospect of climate change has brought despair, chaotic misunderstandings, fear and denial. The denial bewilders some of us, but it is not really surprising: the idea that humans can actually *radically change* planet Earth is, at face value, ridiculous.

It is far easier to look away. But that is no longer an option for anyone who cares about generations to come. In fact, it

is increasingly clear that, indeed, there must be a proportional response to this existential threat. Business as usual is
no longer an option. The denial and reluctance of those most
comfortable with the status quo are understandable, but they
are not condonable. There is simply too much to lose.

Changing people's *minds* – with facts, tables and predictions
– has proven extremely difficult. Even showing people the
miraculous beauty of the planet alongside the predicted
losses is not working. Guilt, anxiety and anger can be moti-
vating forces, but they have debilitating side effects: they are
all soul-destroying.

So I wonder about our *hearts*. Have we ignored our emotional and spiritual connections to the planet? Could the
noise swirling around climate change – science, politics,
media blitzes, as well as the weather disasters themselves –
drown out the voice of a loss so profound that it rests
unnamed in our souls? Could our breaking hearts be part
of the reason we are immobilised?

In the 1950s, Rachel Carson's image of a 'silent spring' due
to the proliferation of pesticides was as heartrending as it
was controversial. Carson was ridiculed, her predictions dismissed. The corporate world paid millions to have her
silenced. But eventually, the love of birdsong won out.
People read Carson's book, grieved at the prospect of a silent
spring devoid of bird- and frog-song, spoke up, and insisted
the chemical-company-supported politicians ban DDT.

The iconic images surrounding climate change are very
stark: the human mother watching her child slowly die from
malnutrition; the majestic polar bear mother with her cub
on a shrinking ice floe; or the head of state of a small island

nation pleading with delegates, at yet another international conference, to save his homeland from disappearing under rising ocean waves.

Earth is a gift rendered in soil, water, atmosphere and living beings. It is a gift we must, by all rational, emotional, ethical and spiritual measures, protect and preserve for future generations. It is a gift to all of us – not just some. We are compelled by the goodness of being human to ensure that those who are most vulnerable are protected from the ravages of a changing climate. If climate change has taken our home, or the life of someone close to us, we feel deep personal grief. But if we are to truly *'love our neighbours as ourselves'* (as universal a dictum as any) we must put ourselves in the shoes of others, and imagine watching *our* malnourished child die, *our* home being flooded away, or burnt by wildfires or covered by desertification. Truly loving our neighbours means feeling empathetic grief.

For many people, the reality of actual climate change remains distant – somewhere else, sometime later. The responses proposed – recycle more, drive less, watch your personal carbon footprint – seem so ridiculously small, futile and incongruous that they are simply dismissed: Why bother? Let me just live my life, right now, the way I always have. Let me enjoy the comforts I worked so hard to achieve.

But I wonder if there could be a hidden sense of deep loss beneath this self-absorbed, preoccupied exterior.

Climate change is one of the most profound experiences faced by humans, as revolutionary a change in perspective as when Copernicus suggested that the Earth moves around the sun. As then, at least for people of faith, the change in our self-image cannot help but tie intimately into our thoughts about God. Some assume that God's second coming will change everything and alleviate the pain and grief from climate change. Others are unsure of how, but are certain that God will not 'let' the human species die. Many of us, though, accept that *'We have met the enemy and he is us'*: Humans are capable of 'free-willing' ourselves into oblivion. That God will remain with us until the end is still a given, but as we are told in Genesis, we are not the only act of creation seen as 'good'. There are no guarantees of human existence on Earth beyond our folly.

It is a world turned upside down.

Decades ago, Joanna Macy spoke about the threat of nuclear annihilation. Her words are tragically appropriate to the reality of climate change today: *'Every generation throughout history lived with the tacit certainty that there would be generations to follow, that its children and children's children would walk the same earth, under the same sky. That certainty is now lost to us. That loss, unmeasured and immeasurable, is the pivotal psychological reality of our time.'* Indeed, we are the first generation of people who now *know* that our children's grandchildren will indeed *not* walk the same earth. They will live on a planet so less hospitable and predictable than it is now that it is unimaginable to us.

Perhaps we are beginning to realise the opposite of Bill McKibben's statement about the *'the end of nature'* from his book of that name. Maybe *nature is ending us.* At the pinnacle of our hubris, we find we are not above the intricacies and workings of the planetary system. The primary loss, then, is of our accustomed relationship to the planet. The associated grief is both deeply personal and hauntingly universal. It is a frightening, *existential grief* that leads to a profound sense of sadness and insecurity.

Therapists and pastors have long recognised that grief is a process. It is to be lived through, not cured. Using the familiar stages of grief defined by Elisabeth Kübler-Ross, climate scientist Dr Steve Running perceptively describes five stages of 'Climate grief': *Denial* that the earth is warming and that the warming is caused by humans; *Anger* that anyone should suggest that their lifestyle be changed; *Bargaining* by suggesting that 'it won't be all bad' (for instance, growing seasons will be extended in some places); *Depression* at the almost unimaginable reality of the problem; and finally, *Acceptance*, enabling active exploration of solutions.

Taking a page from the work of grief counselling expert J. William Worden, coping with the grief of climate change might look something like this:

Accept the loss of our previous relationship to the planet. Denial of climate change is yesterday's news, but we still need to face and name the current and potential losses. Anthropogenic climate change harms the biosphere, threatens modern civilisation and is especially harmful to the poor, already. It is going to be an *'entirely different planet'.* We need to understand what that looks like.

Work through to the pain of our grief. This is the most salient task if we hope to break through the present inertia. The 'dark night of the soul' has always been recognised by traditional paths as sacred. Moving deliberately through the darkest places will help people emerge with empathy and empowerment as they realise their own capacity to change.

Adapt to an environment in which our traditional relationship with the earth has been replaced with a new one. Everything must change, from how we dry our clothes to the underlying assumptions about our economy. We need to support people trying to adjust, trying to make a difference.

Emotionally relegate our old preconceived notions of control and dominance of the planet to the past and move on into a new community of being. Instead of living with guilt, fear and depression, we need to face our profound loss, share our fears and anticipated losses and validate each other's grief. We need to encourage people to actively mourn the changes already here and those to come. We need to dry our tears and embrace a new relationship with Earth infused with wonder, reverence for all life and gratitude.

Blessedly, there is already evidence of a radical response: worldwide concern about the fate of the planet; challenges to basic assumptions about the efficacy of free-market capitalism; and a younger generation, connected planet-wide, that is less complacent about accepting the world of their parents, whether it be a repressive government, a marginalised people, or a food supply produced unnecessarily at the expense of people and the earth. People understand the connections and, increasingly, are suggesting responses that are proportional to the problem.

In the face of my daily despair, I remember the lovely line of geese over our farm as they spiralled their way southward. Maybe they will fly over fields of solar panels; farm fields successfully managed with no ploughing, no herbicides, no fertilisers; cities with 'smart' buildings with green roofs and urban gardens; restored wetlands protecting coastal areas; bicycle lanes and electric cars. If the geese were flying over Europe in November of 2017, they would have seen representatives from 195 nations meeting in Germany to plan for the future of the planet. This was the 23rd meeting of nations to address this planetary threat. Threat: yes. Response: miraculously inspiring, even with its limitations.

Maybe my most heartfelt grief is for my own country. We have so many resources – scientific, technological, economic – that could help the world as it applies itself to preserving a viable future for as many of the inhabitants of the planet as possible. My grief is for the already lost opportunities to help, and for the blindness of our leadership, who do not see that among the world of nations, we are being left behind in the dust, economically, morally and, increasingly, even technologically. The rest of the world adapts to the new reality of the future; we just continue to cry about an unrecoverable past.

'In return for
the privilege of breath'

> *Whatever our gift, we are called to give it*
> *and to dance for the renewal of the world.*
> *In return for the privilege of breath.*

> Robin Wall Kimmerer

In mid-December, when the nights are long and the daytime temperatures stubbornly remain below freezing, high up on the top of the mesas in northeastern Arizona, the Hopi perform the Soyal ceremony – bringing the sun back from its winter journey. It is one of many rituals in an annual cycle, performed in much the same way in the same place for thousands of years.

In college, I majored in anthropology and wrote my senior thesis on the religious beliefs of the Hopi people. I explored how their rituals reflected their seamless relationship with their environment. Attending college in the East, I could use only the secondary sources available to me. But for the summer and fall after my graduation, I spent time in and around the mesas that are home to the Hopi people, working to defend Black Mesa, one of their sacred sites, from strip-mining.

We were not successful. (The fight continues half a century later.)

During my time there I had the unique and blessed opportunity to sit at the feet of a Hopi elder, listening to him describe the Soyal ceremony.

While I do not now remember the details of what the Hopi elder said, I can still hear the timbre of his voice: quiet, 'rez-accented' words flowing out, as if an incantation. And I remember my reaction when he explained the Hopi perform this ceremony each year not just to benefit the tribe, but for the benefit of the whole world.

To myself I said: *Please do not stop, for if you do, the sun might not return for any of us.*

The scientist in me embraces the possibilities of mystical agency inherent in the worldview of indigenous people. Science is limited by what is calculable; this is not the only source of knowledge. Using cause and effect, science seeks to find out *how* things work. But it does not seek more ultimate questions of *why* they should work the way they do. Nor is it theologically heretical for me to affirm the worldview, faith and rituals of religions different from mine. If my Christian faith is open enough to accept the 99 names describing God honoured by Islam (which it is), why should I draw the line at the faith of a Hopi man who bases his relationship to the sacred on many more thousands of years, as compared to our relatively short-lived Abrahamic traditions?

So I can accept more than one explanation to each of the queries of 'how' and 'why' the sun turns back at the winter solstice, even as it is essentially unknowable.

More than anything else, sitting at the feet of the Hopi elder, I sensed a deep truth in the Soyal ceremony, even as I realised it affirmed a relationship so entirely foreign to me (as part of the deaf, dominant culture), as to be incomprehensible.

These forty-five years later, feelings of deep respect are reawakened in my soul by the words and actions of a group of native people.

Standing Rock was a gathering of Lakota Sioux and their supporters in 2016 – 10,000 people at one point – on a windswept prairie in North Dakota. The event turned out to be far more than a protective gesture against an oil pipeline threatening a water supply. Led by the youth from dozens of tribes in consultation with their elders, it tapped into the deep and life-sustaining connections with land and, in particular, water that these and other indigenous peoples from around the planet have sheltered, despite the predominant and often abusive ethos of dominion and extraction from the earth. The 'Water Keepers' at Standing Rock acknowledged and celebrated the sacred connections and, remaining non-violent in the face of state-ordered violence, set an example of a deeply spiritual protest. They were defending their primary water source and a number of sacred sites. That the threat came from a pipeline carrying fossil fuels quickly escalated the events into concerns about climate change. A 'local' issue became inextricably connected to the rest of the globe.

Eventually, Standing Rock gained support from all over the world, demonstrating the power of social media and how it can be used to gather far-flung and diverse strands of deep care for the earth and to knit them into a blanket of protection. That thousands of people responded in support – with donations and by showing up, despite frigid temperatures and less-than-perfect living conditions – is enormously encouraging to me.

That the peaceful witness did not 'stop' the pipeline was of almost less consequence. (The fight continues – in the courts and in politics.)

When a predominately white, European, patriarchal and anthropocentric system, heavily into individualised salvation, arrived on the shores of these lands, environmental knowledge, worldviews and religions of the native people were ignored, or, worse, persecuted.

Times have changed: Indigenous people here in America and around the world are leading us toward a reality they have preserved, often against violent suppression, for millennia. Air, water and lands are more important than profits, they proclaim. Corporations, and politicians in their pay, will not have the last say over something needed and cherished. Fossil fuels have given us much, but at great expense to many people and to land, waters and air. As national governments – liberal and conservative – struggle to remain relevant – there are stories of hope and sanity and compassion at the local community level.

There is deep wisdom, experience and skill offered to the rest of the world by many traditional communities of indigenous peoples. Theirs is a soul-inspiring understanding of the place of humans within the community of earth and I admire the practices that affirm that relationship.

But we must be careful here. Expecting native wisdom to have miraculous answers to problems originating from the dominant society can be cultural exploitation of a different type.

Thankfully, the lines between science and traditional ecological knowledge are beginning to blur. A particularly lyrical description of this blurring process is in *Braiding Sweetgrass: Indigenous Wisdom, Scientific Knowledge and the Teachings of Plants*, by Professor Robin Wall Kimmerer.

Professor Kimmerer, a member of the Potawatomi tribe, teaches environmental biology at the State University of New York in Syracuse. She is the founding director of the Center for Native Peoples and the Environment and part of an active group of academics called 'Rising Voices', which promotes *collaborative science with indigenous knowledge for climate solutions*. Although the indigenous contribution to the causes of climate change has so often been minimal, climate disturbances continue to affect indigenous people first and most severely because of intense poverty on reservation lands and because so many native peoples continue to derive their livelihoods directly from the land and seas. Adaptation is imperative; sharing and collaboration between science and native peoples amplifies understanding and adaptation. Rising Voices offers a community of hope.

There are two aspects of indigenous wisdom that stand out in particular for me, indirectly related to the traditional wisdom/scientific knowledge connection.

One, traditionally attributed to the Iroquois confederacy, is concern for the seventh generation to come. The point is simple but overwhelmingly ignored in our modern world: In all decisions, the generations to come must be considered, even to the seventh. If we take twenty-five years as an average generation – that means *one hundred and seventy-five years*!

From the standpoint of my life, near the end of my sixth decade, grandchildren and great-grandchildren are in the centre of my concern, but add on a century or two and my familial considerations blur. Yet climate change, of all the issues on the planet today, is the one most in need of long-term considerations. 'By 2050 ...' or 'by the end of this century ...' are difficult, if not impossible, for a society built on instant gratification and oil addiction to grasp. Yet this foresight is what the traditional wisdom has advocated all along. It is a bitter irony: climate science forces the dominant culture to think on an indigenous timeline for once ...

The second aspect of indigenous wisdom to stand out for me is the imperative for gratitude and reciprocity. Land is a gift, not a commodity; other beings, plants and animals, are sources of knowledge, not just resources; and humans are only one part of the whole system – always in a reciprocal, symbiotic relationship, not just consumers.

Once again, Robin Wall Kimmerer, most lyrically:

> *Something beyond gratitude is asked of us ... We are bound in a covenant of reciprocity, a pact of mutual responsibility to sustain those who sustain us ... plant breath for animal breath, winter and summer, predator and prey, grass and fire, night and day, living and dying ... Our elders say that ceremony is the way we can remember to remember ... The moral covenant of reciprocity calls us to honour our responsibilities for all we have been given, for all that we have taken. Whatever our gift, we are called to give it and to dance for the renewal of the world. In return for the privilege of breath.*

In the spirit of gratitude and reciprocity, and '*in return for the privilege of breath*', the Hopi, along with many other tribal societies, perform sacred ceremonies '*for the renewal of the world*'.

One would think that a dominant culture, based largely on a book that treats 'covenant' as a sacred agreement and the love of neighbour as an imperative, would understand the *'moral covenant of reciprocity'*. But seemingly, it does not. The gift of the world around us is so taken for granted that even a simple thank you is dismissed as unnecessary. And the concept of 'neighbour' has yet to be extended to rocks, trees, mice and centipedes.

We, of the dominant society, have a blessed opportunity here: to accept our limited temporal and familial understandings of our world. Turning to the renewed witness of indigenous peoples will help us to grow and to heal so that we too can contribute to the renewal of the world. I give thanks to the people of Standing Rock for helping me remember and regain a connection to the wisdom I first experienced at the feet of the Hopi elder.

Choices on the journey

*A thing is right when it tends to preserve the integrity,
stability and beauty of the biotic community.
It is wrong when it tends otherwise.*

Aldo Leopold

Autumn.

The distant snow-sprinkled mountains rise behind the
fading fall hues of the nearer foothills. This afternoon I will
do a final clean up of the garden for the winter, bring in
more firewood, and make a pot of soup.

Ours is a blessed life, and I murmur a prayer of gratitude
for the solace and beauty of it all.

Gratitude is important, maybe even crucial, to spiritual and
psychological wellbeing. But I often feel hounded to do
more, in response to this blessed life. By God? Guilt? Some
might ask, *why?* Why not just enjoy it? A partial, facile
answer might be that the genes and nurture of activist parents compel me always to try to help save the world.

But I think the hounding is more than that.

Action embodies my faith and makes it real. As many wise
observers have noted, religion is something you *do*, not
something you *think* or *believe*. To be faith-full, I need to be
busy, to take the step of trying to embody, yes, the *social*
gospel. I consider myself a Christian because I try hard to
follow the teachings and actions of Jesus, who did justice,
had compassion, healed and cared for neighbours. I do not
believe Jesus' teachings were about personal salvation.
Indeed, I think personal salvation may be an oxymoron. I
also do not believe Christianity is the only active path
toward righteous living, but it is my path at the moment.

It is enhanced by wisdom from other traditions. I find myself deeply moved by Robin Wall Kimmerer's description of a '*moral covenant of reciprocity*'.

Abraham, Moses, David and Jesus had various covenants with God. But why limit the concept to being between a chosen human and God? I see no reason why other members of the created order might not have direct covenantal relationships with God, and certainly I can see the benefits of a covenant of reciprocity with all creatures and plants and indeed with the land itself.

So my personal effort is straightforward: Each day I try to live deliberately, carefully, and as lightly on the earth as I can – being respectful of my neighbours and neighbourhood, which includes the land itself, the soil, plants and animals as well as the human beings, fully realising that I do not always meet the standards that I place upon myself. I think about these things. I write about them because writing helps me work through the thoughts and theology of it all, and because I want to share how living this way brings comfort and meaning to me in the face of a world that is often perplexing and frighteningly out of balance.

On warm-season Saturdays, subject only to the musings of the sun, we hang our laundry outside to dry, watching the mountain peaks to the west. We try to eat local food from the many diverse farms that have sprung up around us. Eating local connects us to the whole community: soil, sun, water, neighbours.

More than a decade ago, my husband and I decided to move to a small community alongside Lake Champlain, which stretches 125 miles northward into Canada, while forming a border between New York and Vermont. To the west are New York's Adirondack Mountains, which give their name to the park that surrounds us. All state land owned within the confines of the park must be kept 'forever wild', as written in a 1894 clause in the New York state constitution. In the early 1970s, the private land holdings within the park came under fairly explicit new conservation regulations. These landscapes are cherished by people who have lived here for many generations, also by summer people who descend in the warmer seasons, and many 'transplants' like us – people who chose to live here full time because of the conserved beauty and wildness of this particular rural environment and community.

The transplants come from a 'somewhere else' that just wasn't quite working for them, many from suburbs and cities, and include a group of young farmers who bless us with a standard of cooperation and community that the rest of us try our best to emulate. We all look out for one another: train to be volunteer fire fighters; contribute vegetables to the food pantry needed by some less fortunate neighbours; help restore houses damaged by floods from hurricanes; and create 'meal-trains' for new parents or families with an ill member. We work to promote community: we restore and volunteer at an old Grange Hall and make it into a magnet for concerts, movies, dances, gatherings; we build trails to provide access to wild places; we support a small alternative school that fosters independence and compassion for others among our youngest members in an outdoor setting of farm and woods.

And, as we go about our daily and weekly routines, we try consciously to increase our awareness, asking ourselves: *how* does the particular purchase I am about to make affect and enhance the larger community beyond the borders of our town, with which I am connected?

Do I really need the latest version of anything, be it a phone, a winter coat, or a cheese slicer? We are bombarded with ads for a better or cheaper version of this or that, discount sales (usually markdowns from already inflated prices) and all the other marketing to consume more, all backed by insidious planned obsolescence. This is blatant manipulation, nurturing an addiction to consume more. And, as has been noted by many, does not make us any more happy.

Alternative choices: reduce, reuse, recycle. Barter, swap, renew. Give away.

Some might consider my choices as futile gestures, arising from personal guilt for having much, or, worse, as pious acts designed to flaunt a holy way of living.

Not so. I learned a long time ago that excessive guilt or piety is a waste of time and energy, not a constructive way of living one's faith. These alternative choices, in themselves, will not be enough to heal the hurting planet and mend the economic and other disparities that plague this world. But for me, as a sojourner on this planet, it is unacceptable to proceed independently with my life, taking no responsibility for my fellow inhabitants, especially since I can *afford* to make carefully considered choices, and many cannot.

In 21st century America, our personal life choices have the potential to alienate us from – or to embrace us into – the fabric of an increasingly more interdependent world. By the simple act of eschewing a clothes dryer, our household's carbon footprint is reduced just a little bit, and, hyperbolic extension notwithstanding, helps our neighbours in Bangladesh, whose homes are *already* awash from rising seas due to global warming. By growing vegetables and sharing with neighbours of limited means, I contribute in a small way to their health and enjoyment of good meals while encouraging local food production.

There is a huge spectrum of doing good by the planet: our job as human beings is to figure out where we want to be on that spectrum. Everyone has different tolerances, concerns and, of course, resources.

I realise I am blessed with the luxury of 'more than enough' to make such choices. What all of us do *not* have is the luxury of ignoring limits to growth and the imperative of making sure everyone has equal access to and the blessings of participation in a sustainable community. How we live our lives well is not about satisfying our personal needs as human beings, for convenience, for cheap goods, for oil, water, food, or even security from terrorists. Rather, it is about our collective needs and responsibilities as co-inhabitants of planet Earth. How we live our lives is not limited to the health of just the particular countries or ecosystems in which we live. It is about the whole earth and climatic system. When a choice is set before us, especially before those of us who have many, many options – to recycle, eat less meat, hang out the laundry, use LED lights, drive the most fuel-efficient vehicle we can afford – can't we eschew the easiest and simplest choices for us as individuals? Do we really treasure our personal preferences and autonomy more than the interwoven bonds that connect us to this green-and-blue-sphere-with-white-swirls set against the black expanse of space?

We cannot now run away from accepting responsibility: there is no 'away' to run to any more. There never was.

By choosing a lower-carbon/community-enhancing path, we embrace and are embraced into the greater *oikos*, or household, of planet Earth.

We are not alone. My husband and I choose to be active Associates in the Iona Community, a dispersed ecumenical community originating in Scotland, with Members and Associates throughout the world sharing a common vision of what the incarnation of community, based on the teachings of Jesus Christ, might look like. We follow a 'rule' – a code of practice and discipline – that includes daily prayers for each other and the world, work for justice and peace within our home communities and beyond, and periodically meeting together. When the world seems inhabited, and indeed ruled, by people lacking compassion, lacking any sense of justice or a moral compass, we find that support from this community is crucial, reminding us that we are not crazy or alone.

All of the choices we make '*in return for the privilege of breath*' become sacred acts, our contribution to the ongoing process of creation, acts of reciprocity, in response to gratitude. For me, these choices are active forms of prayer. The gestures help my neighbours and heal the planet a little.

And they help to mend my soul.

Prophets, end times
and hope

> *It's somewhat comforting knowing that things*
> *are going to fall apart, because it does give us that*
> *opportunity to drastically change things.*

Tim DeChristopher

As a sojourner in northeastern USA, shades of green are the predominate hues painted on the landscapes I love, at least in the summertime. But when I travel to the southwest, the beauty of the expansive intensity of sculpted bare rock in russet brown, tawny yellow and rust, often with no green tree or bush in sight, takes my breath away.

Southern Utah, with its beautiful sandstone landscapes, contains thousands of square miles of federally designated wilderness. It is an area beloved by climate activist Tim DeChristopher, who chose to perform an ingenious and brave act of civil disobedience, in defiance of the sale of oil and gas leases on these lands. He went to a bidding auction and offered $1.8 million on 22,500 acres of land with no intent or means to pay for it. He won the bid, thus temporarily derailing the process. He was arrested and eventually sentenced to two years in jail.

I think about prophets, Jeremiah, Bill McKibben, Tim DeChristopher. Jeremiah spent over forty years pleading with the people of Judah to repent. Death threats and prison were the result. Bill McKibben began voicing his warnings about climate change thirty years ago, first with his landmark book *The End of Nature*. Though he has also been jailed for acts of civil disobedience, McKibben continues to devote his life relentlessly to the cause. Out of prison after 22 months, Tim DeChristopher also continues to speak, cajole, write and testify.

It is not easy being a prophet. No one wants to be told they must *radically* change the way they live, even when it is increasingly clear that the prophets are right:

'*... things are going to fall apart ...*'

Indeed, they already have. Due to the insulating effect of wealth, mobility and technology, many people in 'developed' countries do not realise that they themselves are already experiencing the effects of climate change. And the focused and sophisticated media effort by corporations to attempt to create a different reality – to deny causation, to brand climate activists as unpatriotic, and to lie about known economic and health effects – provides artificially safe cover for those who do not want to change their habits.

Climate change is no longer a threat – it is a reality. So, as we move through the second decade of the 21st century, we face an apocalyptic moment. Rather than surrender to a dystopian future, Tim DeChristopher chooses to see this moment as positive: '*... it does give us that opportunity to drastically change things*'.

There is hope in this statement. But lest we succumb to 'cheap hope', we must consider for a moment the word '*drastically*'.

For civilisation to survive in a manner recognisable to us, the human global community needs a surge of changed perspectives and habits, particularly within the Western 'developed' world. In the titles of their books authors (and undoubtedly their publishers) are not shy about expressing the size of the technological, sociological, economic and, of course, political changes that must take place: *The Great*

Turning, by David Korten. *Everything Must Change*, by Brian McLaren. *This Changes Everything*, by Naomi Klein. All these describe the drastic necessity for a profound rethinking of how humans must now go about their business of living on this planet.

This is serious business. At the moment, humankind continues to display grave misconceptions about the potential for unlimited growth, muting the limits of carrying capacity with an astounding disregard for the quality of life of our neighbours. Let's remember the lack of consideration about future generations that led to the demise of civilizations before us. The altered state of the planet from climate change *does* change everything: As long as the market makes no allowances for externalities such as natural limits to growth, the deteriorating ecological health of the planet is a threat to free-market capitalism, and, ultimately, to civilisation as we know it.

The implications of the present trends, predictions and our attempts at alleviating the harm will not fully play out for a long time; few of us will be around to witness the effects. (We could learn much from the patient faith of the architects and builders of cathedrals during the Middle Ages.)

Radically changed habits will mean sacrifice. I think people *will* sacrifice to protect that which they love; the preciousness of creation could foster gratitude and elicit a willing response. Many people will sacrifice also for the benefit of their neighbour, out of pure compassion, or to cement their own membership in a community that supports them. As the world is shrinking, the notion of neighbourhood is changing, limited only by the expansive boundaries of our compassion. Environmental sustainability and social justice walk hand in hand. Most every thoughtful thing we do to reduce our impact on the planet does eventually also reduce a negative impact on the poor.

Some people of faith have responded to climate change with new commitment. Greening churches for energy efficiency and creating eloquent and beautiful worship services to celebrate connections with creation can be inspiring. Promotion of 'creation care' may motivate some to pick up trash around their town. Special sermons and planting trees on Earth Day are fine. But can't we try to move beyond *one* day per year to think about the earth? Fostering 'stewardship' of the earth may work for some people of faith, but for me, the word implies that humans are still in charge. Despite the present state of the world, this may ultimately turn out to be an illusion.

Everything is still in a state of becoming.

Greening, creation care, Earth 'Day', stewardship, although all well meaning, trivialise the problem. Such actions may alleviate local problems and assuage the guilt of some consciences, but are not expansive enough on their own by a long shot. We face not only a deeply moral question, but an *existential* problem for people and many species right now, and for all our descendants.

There has been a discomforting shift in the popular press: calls for the *prevention* of climate change have lessened, replaced by words such as *mitigation* and *adaptation*, as if we have accepted the inevitable and given up trying to stop or reverse climate change. On the other hand, recognition of the possibilities of mitigation and adaptation may move us forward with purpose, in time.

Most famously echoed by Rev. Martin Luther King Jr, transcendentalist Theodore Parker said: *'The arc of the moral universe is long, but it bends toward justice.'* Naturally, we gravitate toward the second part of this as it gives us hope, but we must not forget the reality of the first part, with the hard word: *'long'*. If we are to prevail as a species, our perspective, patience and perseverance must be long. At least seven generations long, say our indigenous teachers.

There is an intense need for strong moral leadership here to accelerate collaboration. Throughout history, courageous communities of faith have vigilantly challenged social norms when necessary, often at the bidding of prophets. Christian communities have the model of Jesus keeping faith with his God and trusting that with God – all things are possible. Difficult messages can be conveyed and effective responses carried out in these communities.

So the challenge for communities of faith is to embrace a perspective that recognises God's love for absolutely *all* creation, from worm to wallaby; to extend compassion for all human climate-change refugees of today – from those living in the war-inducing, drought-plagued deserts of Syria to residents of the Isle de Jean Charles off the coast of Louisiana, 98% reduced in size by erosion due to higher sea levels – to rise to the moral leadership the prophets demand of us.

We can be encouraged by hopeful signs of positive change: States, cities and even some corporations are setting new standards to reduce emissions and mitigate problems, despite the ridiculously dysfunctional, reactionary and regressive moves on the part of the United States federal government. Twenty years ago I had to search hard for articles about climate change in any public media. These days, I find it hard to keep up with the flow of helpful, persuasive public information. Concerns and solutions to our systemic global problems are now shared by an array of people via the Internet.

As I enter my eighth decade, I find hope enlivened by remembering the resolutions of situations I once thought impossible. I vividly remember the TV images of Nelson Mandela walking free, of young people scaling the Berlin Wall, and how I wept late at night on that Tuesday, November 4, when a black man became President of my country.

Hope is always flowering in the dire words of prophets. Otherwise, wouldn't they just keep silent?

Tim DeChristopher's last statement to the court, as he was taken away to prison, was: *'This is what love looks like.'*

Jesus overturns the tables of the moneylenders in the temple. He touches lepers. He eats with prostitutes. And – he gives up his life for his friends and for the future of the world. *This is what love looks like.*

'*Active hope,*' says Joanna Macy, '*is a practice ... it is something we* do *rather than* have.' Turning off lights, recycling plastic bags, hanging out laundry, attending a protest, writing a letter to the editor, calling government officials, voting in representatives who understand, running for office to make the changes needed. Reading, thinking, listening, trying to embrace new perspectives on the place of humans on this small sphere travelling through space. Talking, praying, if that is something you do, and tenaciously holding on to hope.

This is what love looks like.

Field with a view

My childhood was spent mostly near woods. Sometimes, in need of solace or simply inspiration, I would retreat to a grove of magnificent white pines, sit against their broad trunks, and listen to their soft comforting whispers.

Trees adorn places I still sincerely love. But now, in my middle age, it is the soft rustle of wind-blown tall grasses that beguiles me.

Over ten years ago we moved to a farmhouse in northern New York. The house sits alongside some barns on a back road that crosses a broad expanse of hayfields bordered by woods. Living here has affected my psyche profoundly, and my need for open vistas is now almost an addiction. We go away, sometimes to those tree-adorned places of my youth, but when I return home to this farm, open as it is to the sky, I feel relief, a freeness of existence, a stand-up-straight-and-breathe-in-with-thanks response to the fields.

On daily walks I sometimes skirt the newly wooded pasture areas, remaining in the wide-open fields. But other times I deliberately pass through the woods, thrilling to the heightened anticipation of emerging once again into the fields. The angle of one field view, looking upslope to the farmhouse from a swale below, is similar to Andrew Wyeth's *Christina's World*. I feel thrown into that painted scene; entranced and heady, I turn – surely the artist is there behind me …

I read somewhere that most people, even city dwellers, feel more comfortable in a field than in a forest, a holdover from our ancestors. There is security in being able to see the predator approaching on our flank. In modern times, this translates into an aesthetic (as opposed to a survival-based) preference for the vista. Apparently we prefer it to be accentuated by a tree here or there, presumably available to climb, should lions pursue us. Frederick Law Olmsted surely acknowledged this when he and his colleagues went about designing New York City's Central Park and Boston's Emerald Necklace.

The fields around our farm are human-created and human-maintained. The grasses, accustomed to being grazed by animals, produce a limitless supply of new leaves, and go on growing even when they are eaten, torn up, or cut for hay for off-season use. The argument could be made that we have cultivated an unnatural alliance; the grasses and, eventually, trees should be allowed to return. But how would the meadow mice and the harrier and the coyotes feel about that?

Our house, set as it is in the middle of the fields, is like a wildlife blind, providing an exceptional opportunity simply to observe. Indeed, we spend a good deal of time each day looking out of the windows, reveling in the beauty of each season, each hour.

In winter, small flocks of snow buntings scurry to collect stray seeds alongside the road. Crows flock into the top of a huge surviving elm and then float down one by one, restocking their gizzards with fine mine tailings: garnet and wollastonite, typical of country roads in these parts. Four-footed animals have been wisely wary of crossing the fields during daylight, but our early-morning walks or skis on fresh snowfall reveal the night wanderings of deer, coyote, fox, rabbit, and my favorite, the delicate winding thread-trail of mice.

After the snow melts in early spring, time pauses. There is a quiet waiting for something to happen. The fieldscape surrounding the farmhouse is uniformly yellow-brown, flattened, dull and cold. The northern harrier sometimes returns precociously to take advantage of no-snow-and-not-yet-green-grass. She drifts, about 10 feet above the ground, a solitary vector moving across the fields in an unhurried survey for prey. A flash of her white rump accompanies each gentle adjustment and course correction.

As the season progresses, a succession of painterly hues emerges: lighter apple-greens mature into grass-greens, soon to be sprinkled with canary-yellow dandelions. By early May, the grass is tall enough to welcome the most precious of our neighbours, returning from South America: the bobolinks.

Males reach the fields first, with a song that Peterson's *Field Guide* describes as '*ecstatic and bubbling, starting with low reedy notes and rollicking upward*'. My winter-dulled soul never fails to soar in response to the year's first strains of bobolink song. The males rise into the sky with rapid wingbeat, seemingly desperate to attain a certain altitude above the grasses, and then float earthward, wings quivering slightly, outstretched as to embrace the field itself, the offer of love bubbling from their throats.

In spring, breezes playing through the softness of new grass growth make very little sound. But by midsummer, the long leafy bracts of tall grasses catch the wind with a rustle not unlike waves breaking on a beach. By then, the female bobolinks are nesting quietly, unwilling to bring attention to their whereabouts. The males perch guard on top of the tallest grasses some yards away.

When I venture out during this season, I am careful to keep to the narrow trimmed path we allow ourselves. The dog accompanying me sometimes disappears; all I can see is a slight movement of the seeded grass tops. It could be something stalking me; but I know it is she – seeking an unwary vole. I call her back onto the path and proceed – stretching my arms to both sides, hands turned downward, feeling the tops of the grasses against my palms as I walk. I think of a line of soccer players, touching the palms of their opponents after a game; touching the grasses this way makes me feel young.

By autumn, in parts of the fields that have not been mown for hay, the grasses have fallen over with the wind and the weight of their own biomass. The winds have torn messy swaths, and the fields look like a tablecloth that needs ironing.

Snow will once again smooth things out.

The fields surrounding our farmhouse seem so grounded, steady, predictable. It is hard to imagine how they might be affected by climate change. As is the case with most temperate regions, local studies suggest the fields will have to contend with generally warmer temperatures, more precipitation (snow or rain), and more extreme weather events. So far this winter, it has not been cold enough to snow very much. Last summer it was too wet to hay parts of the fields. Situated as we are on a ridge, our downhill fields were simply saturated by Tropical Storm Irene, while in the valley hamlet just below the fields, families were flooded out and two houses were ultimately lost.

In the moment in which I write, it is late January. The temperature outside is 41 degrees Fahrenheit, but in a couple of days it is due to plummet to the single digits or even below zero. It has been doing this up/down business for most of the weeks since Christmas. As a cross-field skier, I am particularly vexed at the lack of significant snow cover directly out my backyard.

But that which causes my casual annoyance can be a matter of survival for other species. Snow insulates, so even when air temperatures fluctuate, the ground keeps a more steady frozen state. The roots of plants stay secure, and underground living processes slow or cease. Without snow, unseasonal warmth like today will fool some systems into starting up, only to find that they have succumbed to false advertising. The energy expended is thus lost, weakening the organism, if not killing it.

Studies are underway on the effects of these freeze/thaw cycles on fields over long periods of time. Root injury and changes in the timing of nutrient availability and photosynthesis affect the overall growth and health of plants, leading to changes in species distribution. And, of course, opportunistic and invasive species are always ready to move right into the areas vacated by natives.

Temperature, moisture, intensity of weather events: we tend to concentrate on magnitude and forget the effects of temporal shifts. An extra-early warm spring day can bring about a black-fly hatch. A subsequent freeze kills them all. Nice for us, we think – but the birds whose migration is timed to 'catch the hatch' sorely miss the flies. Maybe this one year, it won't matter so much. But over a decade or so, there could be significant alterations in the timing of insect hatches, as well as in seed production, dissemination and germination of some species of grasses and herbs. *Phenology*: the study of the timing of life-cycle events of plants and animals. Results provide more hard evidence of climate change.

Our bobolinks eat seeds, grains and insects. Has anyone told them that someday the menu choices might not be available when they arrive here in late spring?

How will the vole population be affected by a timing shift in seed production? And if the vole population changes, how will that affect the harrier's chances of feeding her young?

My observations from the farmhouse are not passive. Inevitably, I am invited to consider distance. My gaze, unconfined, is pulled straight outward from where I stand until some limit interrupts: a ridgeline, forested hillside, range of mountains. How is it that yesterday the mountains were so close, but today they seem a distant mirage? Can I see the children playing outside the farm preschool perched on the hillside across the valley? If I were a hawk, would I be able to see right into the kitchen window of that dot-of-a-farmhouse at the far edge of the field? Is my seeing limited not just by geological obstructions and the range of my eyesight, but by my imagination as well?

I feel the acute smallness of my being, accentuated by the broadness of the fields surrounding me. I become but a speck in the landscape, like the tiny human element inserted into the Hudson River School paintings. I have to move in close to search for my presence in the immensity of the field-scape that ignores me.

Perhaps I am not even here at all …

The field remains indifferent to me. My view of it is but an instance, a snapshot; it holds a temporal and spatial existence far beyond me. I feel solace, not terror at this. Being an insignificantly small part of it all is grace-filled for me, and my gratitude is beyond measure.

Sources and acknowledgements

No inconsistency

p.18, attributed to Teilhard de Chardin, in *The Luminous Web: Essays on Science and Religion*, by Barbara Brown Taylor, Cowley Publications, Cambridge, MA, 2000, p.100.

p.20, Pope Francis, in *Laudato Si: On Care for Our Common Home*, United States Conference of Catholic Bishops, Washington, DC, 2015, pp.96-97 © Libreia Editrice Vaticana, Vatican City.

Wild apple trees and Process theology

p.24, Alfred North Whitehead, from 'Religion and Science', *The Atlantic*, August 1925.

pp.29-30, Bruce Epperly, © Bruce Epperly, 2001, *Process Theology: A Guide for the Perplexed*, T&T Clark International, an imprint of Bloomsbury Publishing Plc, 2011, pp.70, 71. Used with permission of Bloomsbury Publishing Plc.

Imago Dei

p.32, William Blake, from 'Auguries of Innocence'.

Earth excepting humans?

p.38, Aldo Leopold, from *A Sand County Almanac and Sketches Here and There*, Oxford University Press, 1968, p.110, © 1949, 1977 by Oxford University Press, Inc. Used by permission of Oxford University Press.

p.41, Paul J. Crutzen and Christian Schwägerl, in *Living in the Anthropocene: Toward a New Global Ethos*, in Yale Environment 360 (online magazine), January 24, 2011. Used with permission of the magazine.

p.44, Mary-Jayne Rust, in *Climate on the Couch: Unconscious Processes in Relation to Our Environmental Crisis*, Annual lecture for Guild of Psychotherapists, November 17, 2007, pp.6-7, © John Wiley and Sons. Used with permission of John Wiley and Sons.

p.45, Bruce Epperly, © Bruce Epperly, 2001, *Process Theology: A Guide for the Perplexed*, T&T Clark International, an imprint of Bloomsbury Publishing Plc, 2011, p.98. Used with permission of Bloomsbury Publishing Plc.

Grief

p.52, Joanna Macy, from 'Planetary Perils and Psychological Responses', in *Psychology and Social Responsibility: Facing Global Challenges*, edited by Paula Green and Sylvia Staub, New York University Press, 1992, pp.6-7. Used by permission of New York University Press.

p.53, Steve Running, *5 Stages of Climate Grief*, blog entry on Friends of 2 Rivers website (www.friendsof2rivers.org), 2007. Used with permission of Steven W. Running.

A version of the chapter 'Grief' appeared in *Sojourners* magazine, August 2013.

'In return for the privilege of breath'

Choices on the journey

Prophets, end times and hope

Field with a view

Portions of this chapter and the photograph on the cover of the book appeared before in *Places We Live*, an online publication of *Orion Magazine*, https://orionmagazine.org.

Wild Goose Publications is part of the Iona Community:

- An ecumenical movement of men and women from different walks of life and different traditions in the Christian church
- Committed to the gospel of Jesus Christ, and to following where that leads, even into the unknown
- Engaged together, and with people of goodwill across the world, in acting, reflecting and praying for justice, peace and the integrity of creation
- Convinced that the inclusive community we seek must be embodied in the community we practise

Together with our staff, we are responsible for:

- Our islands residential centres of Iona Abbey, the MacLeod Centre on Iona, and Camas Adventure Centre on the Ross of Mull

and in Glasgow:

- The administration of the Community
- Our work with young people
- Our publishing house, Wild Goose Publications
- Our association in the revitalising of worship with the Wild Goose Resource Group

www.ionabooks.com

The Iona Community was founded in Glasgow in 1938 by George MacLeod, minister, visionary and prophetic witness for peace, in the context of the poverty and despair of the Depression. Its original task of rebuilding the monastic ruins of Iona Abbey became a sign of hopeful rebuilding of community in Scotland and beyond. Today, we are about 250 Members, mostly in Britain, and 1500 Associate Members, with 1400 Friends worldwide. Together and apart, 'we follow the light we have, and pray for more light'.

For information on the Iona Community contact:
The Iona Community, 21 Carlton Court, Glasgow G5 9JP, UK.
Phone: 0141 429 7281
e-mail: admin@iona.org.uk; web: www.iona.org.uk

For enquiries about visiting Iona, please contact:
Iona Abbey, Isle of Iona, Argyll PA76 6SN, UK. Phone: 01681 700404
e-mail: enquiries@iona.org.uk